HELICOPTERS

Chris Chant

GALLEY PRESS

CANADA

HELICOPTERS

Chris Chant

Published by Roydon Publishing Co. Ltd.
London
England

Published in this edition by Galley Press, an imprint of
W. H. Smith and Son Limited,
Registered No. 237811 England.

ISBN 086136 8010

Printed in Italy

Produced by Talos Books Limited
Lincolnshire
England

Edited by Diane Moore
Designed by Brian Benson
Picture Research by Jonathan Moore

Contents

Chapter 1
An Unflying Start

First to grasp the implications of rotary-wing flight with a fair degree of completeness was Leonardo da Vinci, illustrated being his model for a helical-screw model, developed in 1486-90. It was a depiction of this model that inspired Igor Sikorsky to take up helicopter design.

From virtually the moment that men began to dream about flying in the fashion achieved so perfectly by the birds, it was clear that there were two ways to go about the task: to take-off after a running start, in the manner of most waterbirds, or to take-off vertically from the ground in the manner of the smaller land-based birds. The dreamers of the period before 1500 AD could and did not come to grips with the technical and aerodynamic problems associated with the two concepts of take-off, fixing their attention more romantically but less practically on attempts to mimic the wing forms and wing motions of the birds as a means towards flight.

Serious thought about flight began with Leonardo da Vinci, that vitally important 15th century Italian genius whose talents paved the way for progress in so many fields. And while da Vinci failed to realize the full futility of slavish copying of bird flight in any effort to raise man into the air, he did come to see that machines with fixed wings would require wheeled landing gear to permit a take-off run, and that an alternative was the helical-wing aircraft that could rise and descend vertically without the need for a take-off run.

Thus was born the concept of what is today called the helicopter, the word being derived from the Greek *helix* (spiral) and *pteron* (wing). It recognized that fact that to generate lift a wing of any sort must have air moving round it. This movement can be obtained in one of two ways: if the wing is fixed relative to the body of the whole machine, then the aircraft must move through the air fast enough for the pressure differential in the air moving faster over the upper surface and slower over the under surface to generate a lift more than equal to the weight of the whole machine; alternatively, if the wing can move in relation to the body of the whole machine, then the aircraft as an entity can remain stationary while the moving wings generate the lift. This latter concept is used in the helicopter, in which narrow wings called rotor blades are arranged round a central shaft (like spokes on a hub) which is powered so that the rotor blades are swept through the air to generate lift and so raise the whole machine vertically.

The notion is the mirror image of what happens in the windmill: here the wind strikes the blades, which turn and drive the windmill's grinding mechanism. The helicopter is like a powered windmill laid on its back, with the blades powered to push the rotor mechanism and its attachments straight up into the air. The world's first helicopters were called *moulinets à noix,* the term taken from children's toys, apparently popular in the first half of the 15th century, and consisting of spindles with feathers attached at 90° angles to each other and with their leading edges slightly raised: when the spindle was held in the hand and 'spun up' with a wound string, the whole device rotated, generated lift and flew up into the air. Evidence for such toys, still common today, can be traced to as far back as 1325.

Given his fascination with all things to do with flight, it is not surprising to find that da Vinci developed sound though limited ideas about the helicopter, a design dating from 1486-90 showing a screw-like that could have lifted itself into the air given the application of sufficient power. However, da Vinci thought in terms of man power, which would have been totally inadequate, and the device also lacked any form of control.

There followed a gap of some 250 years before the basic notion of da Vinci's helicopter was revived by the French mathematician Paucton in his Ptérophore design of 1768, which remained unbuilt but is still of interest as it featured a rotor for lift and a propeller for propulsion. More important by far was the model built in 1784 by Launoy and Bienvenu, and flown successfully before the Académie des Sciences in Paris. The model, the first such device to be thought out scientifically and tested successfully, had two counter-rotating rotors, each made of wire and covered in silk, driven by a bow-string device. The success of the Launoy and Bienvenu model created a deal of interest, and copies of the device were made on what were by the standards of the day a quite extensive basis. One of these was later seen by the pioneer British aeronautical thinker Sir George Cayley in 1796, and his fascination with the model urged Cayley into a a vastly important career, during the course of which he invented the configuration of the modern aeroplane.

Cayley's first attempts to develop a model helicopter reverted to the *moulinet à noix* configuration, though the bow-string mechanism drove counter-rotating rotors at the top and bottom of the spindle. This model was perfected in 1809, but Cayley then turned his attentions to the fixed-wing aircraft, returning to the helicopter only in 1853, when he published the design of an Aerial Carriage with a wheeled body and fixed tailplane, two counter-rotating rotor units, one on each side of the body, and two pusher propellers for propulsion. Cayley's last flying model, produced in 1853 just four years before his death in 1857, was for a single-rotor helicopter with three tin blades attached to a wooden spindle accommodated in a hollow tube and spun up with a wound string. According to Cayley, the model would rise without difficulty 90 ft up into the air.

By the second half of the 19th century the new science of aeronautics was well into the blood of adventurous designers and thinkers, and notions for fixed and rotary-wing aircraft abounded, their

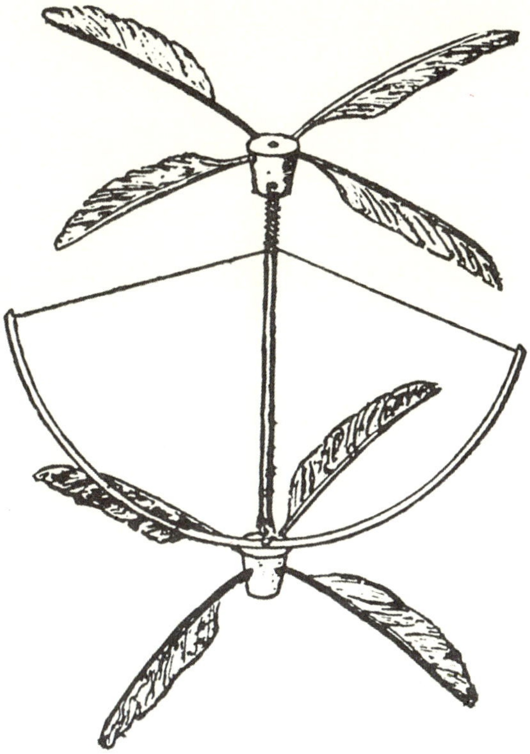

Above: Working on the notion pioneered by Launoy and Bienvenu in 1784, Cayley in 1796 produced a successful helicopter model based on a bowstring mechanism to drive the two contra-rotating rotors, each consisting of feathers inserted at a slight angle of incidence into a cork.
***Below**: Cayley's 1843 design for a convertiplane was phenomenally advanced in concept; the rotors, for example, could be closed down to form circular wings for 'conventional' propeller-driven wingborne flight.*

most common denominator being an excess of enthusiasm and a marked absence of practical realism. Amid this upsurge of mechanical creativity, a few designs stand out for some particular feature: in 1859 an Englishman, Henry Bright, received a patent for a twin-rotor design that has become a standard for such machines today, with contra-rotating rotors on a single drive shaft; slightly later the Vicomte de Ponton d'Amecourt perfected in France a model with contra-rotating rotors driven by a small steam engine; further advances were made by the Frenchman, Penaud, in 1870; in 1874 the German engineer, Achenbach, made a vital contribution, though it was little appreciated at the time, with the development of a small rotor to counteract the torque of the main rotor; and in 1877 the Italian Enrico Forlanini built and flew a steam-powered model with contra-rotating rotors, which rose to a height of some 40 ft and remained airborne for about 20 seconds.

It will be noted that much attention was devoted to twin-rotor designs, and this was because the earliest designers soon came across the problem of torque: put as simply as possible, this means that the action of the rotor turning in one direction causes the body of the machine to react in pure Newtonian terms by turning in the opposite direction, with dire consequences on the controllability of the whole machine. The solution most commonly adopted by the pioneers was the use of twin rotors, turning in opposite directions so that each cancelled out the other's torque. But this can lead to a cumbersome machine with heavy rotors

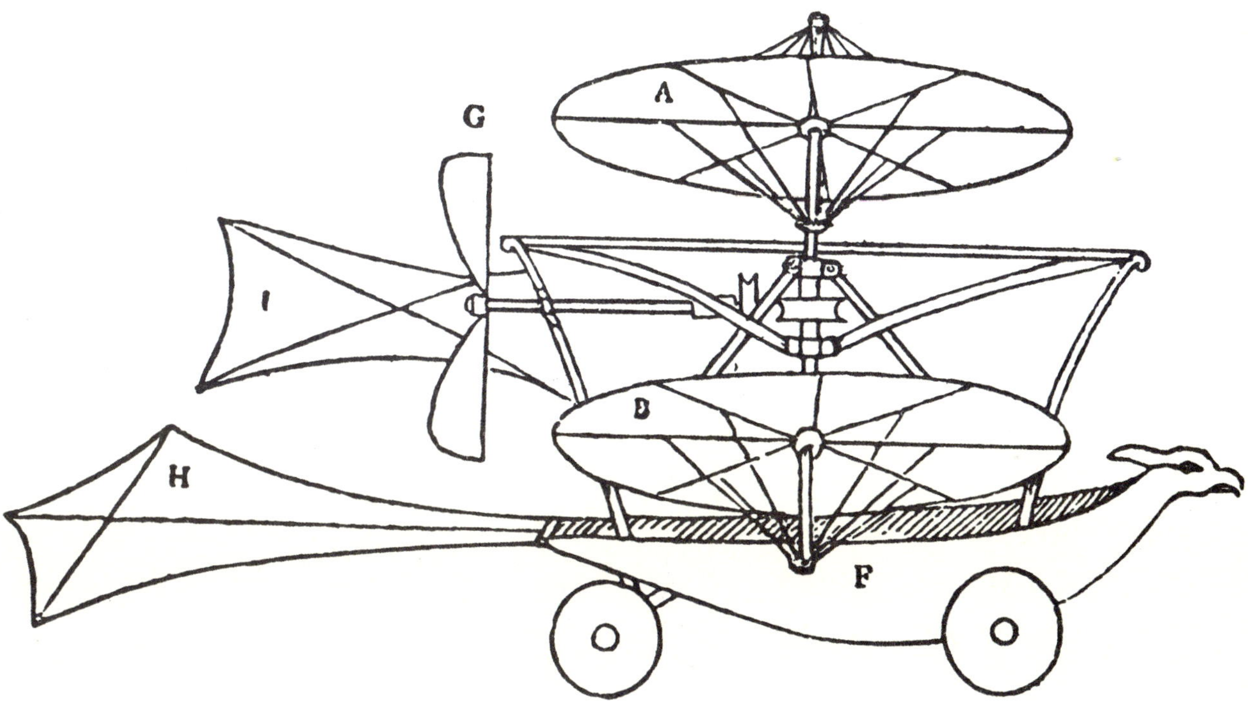

and heavy drive mechanisms, so considerable thought was given to the way in which the torque of a single main rotor might be cancelled: Achenbach found the answer, with a small rotor aligned vertically on the fuselage of the machine some way behind the main rotor to provide thrust to cancel the torque. Unfortunately for the development of the helicopter, Achenbach's work was little known at the time and soon forgotten.

The last quarter of the 19th century was marked by a single-minded devotion to fixed-wing aircraft, and it was not until 1907 that the first real helicopter was built and flown. This was a massive machine designed by Louis Breguet and Professor Richet, and built in France as a four-rotor helicopter powered by a 45-hp Antoinette engine. Decidedly impressive in action but strictly limited in performance, the Breguet-Richet helicopter did manage to lift itself and its pilot to a height of about 5 ft, but lacked any means of control and had to be steadied in flight by a man at each of the four corners. Thus the Breguet-Richet machine was a dead end so far as further development was concerned, but did prove that the vertical take-off aircraft was possible.

In the same year another French machine, designed by Paul Cornu, also flew. Like the Breguet-Richet machine it was powered by an

***Right and below**: The Breguet-Richet No. 2 Gyroplane was produced in 1908 with two sets of two-blade rotors and powered by a 55-hp Renault engine. This made several successful 'hops' (right), before being transformed into the less successful No. 2bis of 1909. Note the large wings.*

Antoinette engine, this time of 24 hp, but was a twin-rotor machine with booms extending from a central framework like the chassis of an outsize pram to support the two two-blade rotors. The Cornu helicopter first took to the air in November 1907, ascending just a few feet into the air, but like the Breguet-Richet machine a design dead end as it lacked any means of control.

Further work was undertaken by a number of designers operating almost entirely independently of each other, an example being the helicopter designed and flown in 1912 by the Dane Jacob Ellehammer: this was of advanced concept, with twin rotors each consisting of a large-diameter ring with six projecting vanes. The rotors were driven by a small piston engine which also drove a small propeller for propulsion, but the most important feature of the Ellehammer design was the adoption of a variable-incidence arrangement for the rotor vanes (now called cyclic-pitch control), permitting an alteration of the angle of lift, so permitting the helicopter to be manoeuvred in the direction to which the rotors were angled.

World War I then intervened to halt the mainstream of development, though it should be noted that one of the great pioneers of helicopter design, the Austro-Hungarian Theodore von Karman, produced one of the most far-sighted rotary-wing aircraft ever conceived as a direct result of this climactic conflict. For the use of the Austro-Hungarian artillery arm von Karman developed a tethered rotary-wing flying observation platform, which could hover at a height of about 150 ft for one hour. Such machines are making a reappearance in the 1980s, though they might be vulnerable to destruction by ground fire, but offer a useful and relatively safe means of observation by means of a TV camera.

For a purely technical point of view, World War I was important to aviation as a forcing ground for the whole spectrum of features that make up any aircraft: structure, powerplant, aerodynamics, materials and the like. Thus the four years of World War I saw a phenomenally rapid development in fixed-wing aircraft, the machine of 1914 that was little more than a very limited and unsafe aerial runabout emerging in 1918 as a reliable, structurally sound and relatively high performance machine able, in various guises, to undertake a variety of tasks. This technical and design competence soon had spin-offs in the field of rotary-wing flight, and the 1920s were notable for significant advances in the USA, France and The Netherlands.

In the USA the two key figures were G. de Bothezat and Henry Berliner. De Bothezat produced in 1923 a substantial four-rotor design weighing about 4,000 lb, and this was capable of lifting a pilot and up to three passengers, though

no adequate means of control were fitted and so the machine remained little more than an experimental rig. Berliner produced several interesting designs, the most successful of which was evolved with the fuselage of a Nieuport fighter, the wings being replaced by a pair of rotors, one on each side of the fuselage and powered by the rotary engine in the nose of the fuselage.

France was the scene for the work of Paul Oehmichen and the later efforts of Raoul Pescara, who started work on helicopters in Spain during 1919. Oehmichen's work is historically important as his second helicopter, powered by a 180-hp Gnome rotary engine, achieved the world's first officially observed helicopter flight over a 1-km closed-circuit course on 4 May 1924. The Oehmichen machine had an X-configuration, with a lift rotor at the tip of each arm and a total of eight propellers, all powered by the one engine. Of the eight propellers five were disposed horizontally for lateral control, two were pushers for propulsion and the last was in the nose of the machine for directional control.

More important by far, however, was the contribution of Pescara, whose third helicopter was completed in 1924. This significant machine was powered by a 180-hp Hispano-Suiza inline

Most of the problems associated with rotor control were solved in the Cierva autogiro with unpowered rotor, seen here in the form of an Avro Rota Mk I.

engine, which drove two contra-rotating co-axial rotors, each with four biplane blade sets, through a special clutch. This machine managed flights of some 800 yards, and could stay in the air for some 10 minutes. But quite apart from the considerable success of his machine in the air, Pescara should best be remembered as the designer who first combined two of the three basic control systems needed in any real helicopter, namely the cyclic pitch mechanism developed by Ellehammer for directional control, and a new collective pitch mechanism that altered the angle of incidence of all the rotor blades simultaneously to control the amount of lift generated by the rotors, and so dictate the speed with which the helicopter rose or descended. Despite these vital successes, Pescara was unable to progress past the level of his third machine and retired in 1925.

The year 1925 saw the emergence in The Netherlands of the first helicopter to adopt what is now regarded as the standard single-rotor configuration. This machine was developed by von Baumhauer, and was characterized by a single main rotor and small tail rotor at the rear of the fuselage for torque control. Unfortunately von Baumhauer had not fully appreciated the need to synchronize the efforts of the main and tail rotors with an absolute precision, so his machine had separate engines for the two rotors, which meant that the machine was seriously damaged during the early part of the test programme.

Oddly enough, the real breakthrough in design came not with the helicopter, which had reached something of an impasse, but with an allied rotorcraft, the autogiro invented by Juan de la Cierva. (It should be noted, incidentally, that the word autogiro is used only for Cierva designs, comparable machines from other drawing boards being known as autogyros.) Appalled by the fatal crash of a commercial fixed-wing aircraft, which stalled after take-off, de la Cierva set to work to design an aircraft-lifting surface that would not stall if the aircraft lost speed through the air. The result was the autogiro, which used a freely-turning rotor in place of the conventional fixed wing. De la Cierva's indisputable reasoning was that the only way to ensure that there was always a relative flow of air over the wings was to make these wings rotate. The Spanish designer considered the helicopter, but soon apreciated that the type still faced a host of technical and aerodynamic problems, resulting largely from the use of a powered rotor and its attendant torque problems. So was born the autogiro with its unpowered and therefore torqueless rotor: to this extent, therefore, the autogiro was a real rotary-wing aircraft, with a conventional fuselage and tractor engine supported by a freely turning rotor which acted as an unstallable wing. So far as the development of helicopters was concerned, the importance of the autogiro developed after painstaking effort by de la Cierva lay in the articulated attachment between the rotor blades and the hub. In earlier helicopter designs a fixed attachment had been used, with the result that the advancing blades generated more lift than the retreating blades, giving the helicopter a tendency to roll. The articulated attachment devised by de la Cierva allowed the advancing blade to rise, so reducing its effective angle of incidence and hence lift, and the retreating blade to drop, so increasing its effective angle of incidence and hence lift. The use of articulated hinges thus produced a stable rotor disc without tendency towards rolling thanks to the balancing of the lift forces. Ironically, de la Cierva was killed in an air accident during 1936 when the airliner in which he was travelling stalled and crashed, and with its inventor's death active development of the autogiro and autogyro ceased until a revival for purely sporting purposes in the 1960s.

The effect of the Cierva type of flapping hinge was soon felt in the development of helicopters during the 1930s. In Italy d'Ascanio produced a twin-rotor co-axial design with horizontal flapping hinges and ball bearings for modification to pitch – an indication of the degree to which helicopter development had lagged was produced by the

Right and below: *Louis Breguet returned to the rotary-wing field in 1935 with the moderately successful Breguet-Dorand Gyroplane Laboratoire, powered by a 350-hp Hispano.*

world records established by the d'Ascanio machine in 1930, including an endurance of 9 minutes and a hover of 1 minute 30 seconds after a vertical climb to 20 ft; two years earlier de la Cierva had made a flight from Croydon to Le Bourget outside Paris in a C.8L autogiro, the flight including the first rotary-wing crossing of the English Channel.

During the mid-1930s the pace of helicopter developments increased, with important successes in France and Germany. During 1935 the great French pioneer, Louis Breguet, re-entered the rotary-wing field, designing with Dorand the Breguet-Dorand Gyroplane Laboratoire, a substantial co-axial contra-rotating twin-rotor design with flapping hinges and powered by a 350-hp Hispano-Suiza engine. This machine provided a considerable quantity of experimental data as well as establishing a number of records, but was lost in an accident just before being totally overshadowed by the Focke-Achgelis Fw 61 designed by Heinrich Focke and Gerd Achgelis. This took to the air with great panache, and in 1937 and 1938 proved conclusively that the helicopter could succeed.

Based on a conventional fuselage and lifted by two large rotors mounted at the ends of long outriggers, the Fw 61 was capable of forward flight at 75 mph, could climb to 8,000 ft and had a range of about 150 miles. Just as importantly, the Fw 61 proved that even a large helicopter could manoeuvre with great precision thanks to the use of fully-articulated rotors: in February 1938 the great test pilot Hanna Reitsch flew the Fw 61 inside the Deutschland Halle, which measured 250 ft by 100 ft. Further development of the design led to the Fa 223 Drache, the world's first operational helicopter, which was ordered in large numbers during World War II, but built only in relatively modest numbers as a result of priorities afforded first to bomber and then to fighter production. However,

Above: *Though workable in terms of performance and payload, the Focke-Achgelis Fa 223 Drache was penalized by excessive structure weight, and suffered tactically from the limitations imposed by the considerable overall span of the outriggered main rotor assemblies.*
Right: *One of the first (and also one of the longest-lived) light helicopter designs has been the Hiller UH-12, also known as the Model 360 and, to the military, as the H-23 Raven. The type has given excellent service in civil and military applications.*
Below: *The Flettner Fl 282 Kolibri (humming bird) was the first truly successful application of the intermeshing twin-rotor concept for moderate overall width with the practical advantages of the twin-rotor layout, and was designed for shipborne operation.*

the Fa 223s proved their worth in fields such as resupply, mountain troops being kept provisioned and ammunitioned by small numbers of Fa 223s dedicated to the task. Comparable work was undertaken in Germany by Anton Flettner, who saw that the considerable span of the Focke-Achgelis concept would prevent full tactical use of the helicopter. Flettner's concept adhered to the twin-rotor design, but dispensed with the outriggers and located the two drive shafts close together and angled out slightly so that the carefully synchronized rotors intermeshed without the blades' touching. Fully developed as the Fl 282 Kolibri, the Flettner helicopter proved entirely practical, and was used with some success as a shipboard reconnaissance platform during the later stages of World War II.

CYCLOPS

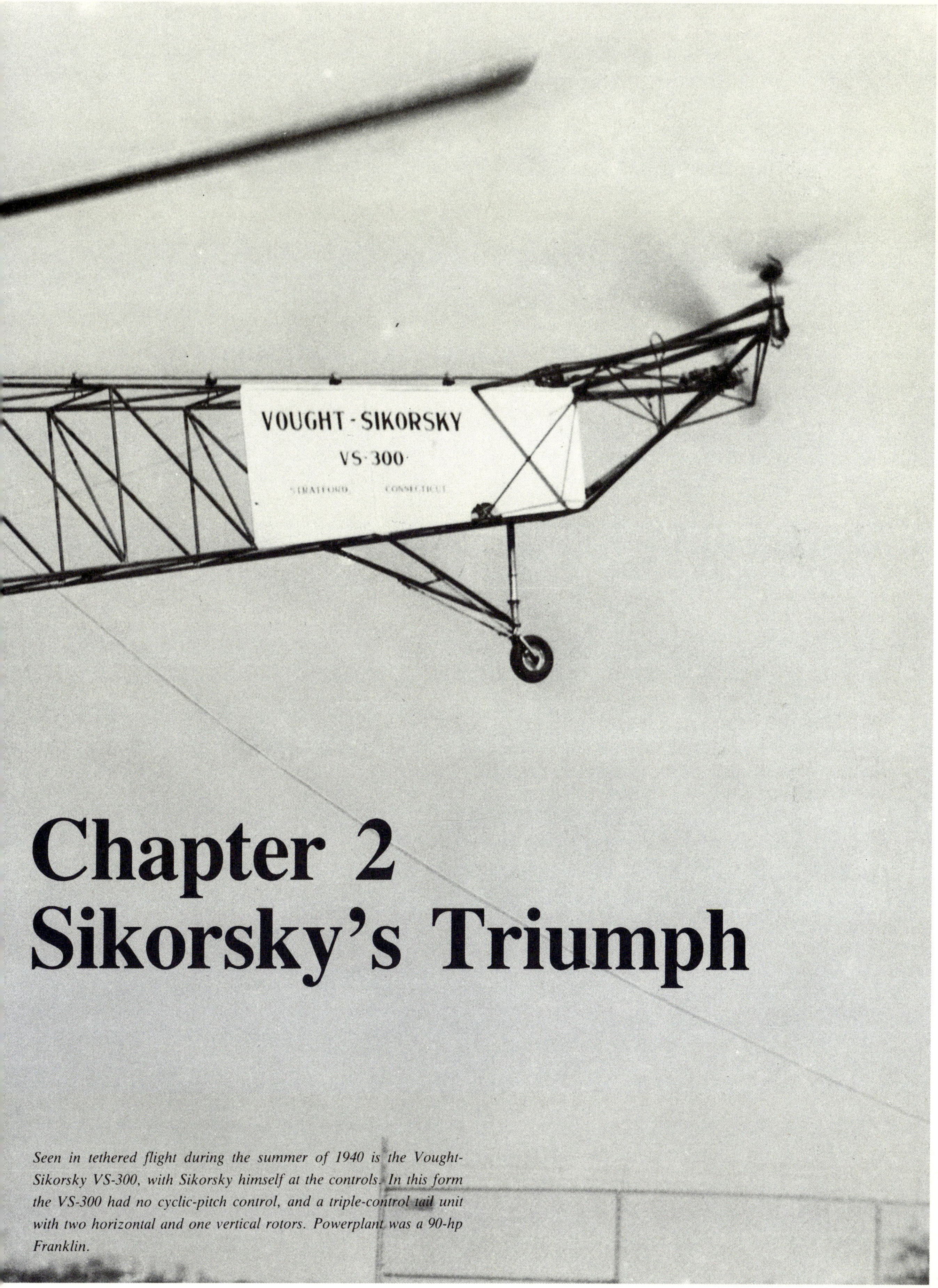

Chapter 2
Sikorsky's Triumph

Seen in tethered flight during the summer of 1940 is the Vought-Sikorsky VS-300, with Sikorsky himself at the controls. In this form the VS-300 had no cyclic-pitch control, and a triple-control tail unit with two horizontal and one vertical rotors. Powerplant was a 90-hp Franklin.

Though men such as Breguet, Flettner and Focke were in the late 1930s coming close to the development of a practical helicopter, they were succeeding to a certain extent by circumventing the problem of torque reaction, opting for a two-rotor configuration (with all its spin-off problems associated with size, weight and complexity) to avoid the unresolved difficulty with the potentially superior single-rotor design.

So it was left to Igor Sikorsky, a naturalized American of Russian origins, to become the true father of the helicopter despite the considerable successes of the European designers. Sikorsky already had an impeccable aeronautical background, having designed the world's first four-engined aircraft (the *Bolshoi*) in 1912 before leaving Russia for France in 1919 and for the USA in 1920. In the USA Sikorsky was responsible for a number of classic flying-boats and amphibians. However, Sikorsky had long harboured a fascination with helicopters, building his first two but unsuccessful designs in Russia during 1909 and 1910. However, Sikorsky appreciated with realistic acumen that a good living was first essential, and he devoted his energies in the USA mainly to commercial products, though a number of patents in the 1920s and 1930s testify to the designer's continued interest in rotary-wing flight.

By the late 1930s Skiorsky felt himself well enough established to return to his first love, where considerable progress was being made by the Europeans as noted in the previous chapter. The result was an experimental type designated Vought Sikorsky VS-300, which may be regarded as the first successful single-rotor helicopter. This machine was based on an open fuselage of tubular construction, perched on a tailwheel type of landing gear with single main wheels and a tailskid, with a single 75-hp Lycoming flat-four piston engine to drive the three-blade main rotor and a small anti-torque rotor at the end of the enclosed tail boom.

One of the keys to Sikorsky's success was the method he brought to his experimental programme. For example, when first lifted off the ground on 14 September 1939, the VS-300 was tethered to prevent the machine going out of

***Right**: One of the first medevac experiments was conducted with a Sikorsky R-6 in 1944, the type then undertaking operation evacuations in areas as different climatically and geographically as Burma and Alaska.*

***Left**: In its original form the VS-300 had cyclic-pitch control, a narrow tailboom supporting a single anti-torque rotor and a large ventral fin. The powerplant was a 75-hp Avco Lycoming engine, and the machine soon showed the need for modifications.*

control and so crashing, and had weights suspended under the fuselage to add a measure of pendulum stability to the design. Full cyclic-pitch control was included in the design, but proved inefficient and was soon discarded.

Seriously unsatisfied by the results of early tests, Sikorsky reworked the design considerably by the time of its first free flight on 13 May 1940: the steel-tube fuselage was now fully uncovered, the 75-hp engine had been replaced by a 90-hp Franklin unit, and the tail had sprouted outrigger surfaces so that the anti-torque rotor could be supplemented by a pair of horizontally-turning rotors for better lateral control. The fact that Sikorsky was on the right basic lines was indicated by the securing of a world endurance record, the time of 1 hour 32 minutes 26 seconds beating the Fw 61's mark by a handsome 11 minutes 37 seconds. A number of modifications were made during 1940 and 1941 before the VS-300 reached its definitive form with a 150-hp Franklin engine, tricycle landing gear, fully effective cyclic-pitch control mechanism and single anti-torque rotor mounted on a short pylon above the tail. The type was tested on water with the aid of flotation bags, and was used for a number of experimental projects right through 1942 before passing to the Ford Museum in Dearborn, Michigan, where the type retains an immensely honourable place to this day as the world's first practical helicopter with a single main rotor. In this context the word practical may be taken to mean that that the VS-300 in its final form was capable of all the maneouvres now taken for granted on conventional helicopters, such as vertical take-off and landing, vertical climb and descent, hovering, and flight forwards, backwards and sideways all under full control.

Just as importantly, Sikorsky had shown that the helicopter was not a freakish flying machine, but an aerial vehicle that was reliable and also useful, and to this extent he may be regarded as the true father of today's worldwide helicopter industry.

By the time that the VS-300 was flown into retirement, Sikorsky was well advanced with the development of the world's first single-rotor helicopter designed for volume production. This was the VS-316A, derived from the VS-300 in basic concept, but re-engineered for greater utility and crew comfort. In basic structure the VS-316A, ordered for the US Army Air Force under the designation R-4, resembled the VS-300, being constructed of steel tube with a covering of fabric. But the main rotor was larger than that of the VS-300, the powerplant was a 165-hp Warner radial, and the two-man crew was accommodated in an enclosed cockpit. In its eventual production form the R-4 had a 200-hp engine, and was used for service evaluation of this new type of aircraft. Aerial observation for artillery and warships was proved feasible, but the type appeared most attractive when used for casualty evacuation in areas such as Burma and the Aleutian Islands, where the climate and geography made the establishment of airstrips difficult, and also made the rapid treatment of casualties imperative. The R-4s also proved the feasibility of air-sea rescue of survivors lying in the water, and the useful capability for communication and liaison work.

The type was also refined considerably in a derivative model, the VS-316B ordered by the military as the R-6 in 1943. This retained the rotor and transmission system of the R-4, but was powered by a 225-hp Lycoming horizontally-opposed engine, and was given a streamlined fuselage with an extensively glazed and bulbous nose tapering into a slim tail boom for an overall shape resembling that of a tadpole. Further evaluation of helicopter roles was again the primary task of these elegant little helicopters, and air-sea rescue rapidly became the type's forte.

But already the success of the VS-300 had spurred other American designers into the helicopter field, the two most important being Bell and Hiller. Bell was already a respected manufacturer of fixed-wing aircraft, with a reputation for innovative designs such as the Airacobra fighter with the engine behind the pilot for greater manoeuvrability and heavy nose-mounted firepower, and the Airacuda escort and bomber-destroyer with wing-mounted nacelles for the gunners. In 1943 Bell flew the first of its Model 30 experimental helicopters, trim little two-seaters powered by a 165-hp Franklin engine. But whereas Sikorsky concentrated on a main rotor with three narrow-chord blades, Bell opted for a two-blade configuration, with wider-chord blades and a tip-weighted stabilising bar indexed at 90° to the blades. This Model 30 was rapidly developed into the classic Model 47, one of the world's most enduring aircraft designs. This type first flew in December 1945, and in March 1946 received the first civil certification ever awarded to a helicop-

The role in which the helicopter has made the greatest impact on the public imagination is that of safety in many of its diverse forms. Air-sea rescue is an important aspect of this role, but perhaps more valuable still is the evacuation of wounded troops from the battlefield, ***Below:*** *A wounded American soldier is moved up to a cleared hilltop for evacuation in a Bell H-13B light helicopter, which has a pair of semi-enclosed litters on the landing gear legs.*

Below right: *Casualties are shifted into hospital from a Sikorsky R-5 during the Korean War, where the medevac concept was fully proved.*

The classic medevac helicopter of the 1950s was the Bell H-13G, seen here with two panniers.

ter. The type entered large-scale production in 1946, and remained in volume production with Bell or with its licensees until 1975 in a host of revised and more powerful forms. The type was also adopted for military service with the US and other forces, resulting in names and designations such as H-13 and Sioux. Payload was initially small, suiting the type for roles such as observation, traffic control and liaison, but the adoption of more powerful engines raised the payload to the extent that the Model 47 in its developed forms could carry an effective external load such as litters for battlefield casualties, a moderate load slung under the fuselage for transport, a spray tank and spray bars for agricultural use, and even (in the military field) light anti-tank missiles or anti-submarine torpedoes.

A competitor in the light-weight helicopter field was the Hiller Model 12. Stanley Hiller has entered the helicopter field in 1944 with the twin-rotor XH-44, designed when he was only 18. But dissatisfied with the extra weight, complexity and size inevitable with such a design, Hiller swiftly opted for the single-rotor configuration in a design known as the Hiller Model 360 or United Helicopters UH-5. As originally flown, this prototype had seating for three in a fully enclosed fuselage, and was powered by a 178-hp Franklin engine. Hiller decided that the payload could be increased, performance improved and costs reduced by a simplification of the design, which entered production in 1948 as the UH-12 with an open, forward fuselage but enclosed tail boom. Successive improvements were added in a variety of models produced throughout the 1950s, when the type was also widely used as a trainer by the US services. Extra power was provided, and among the fuselage options were car-type or 'gold-fish bowl' cabins for varying numbers of passengers.

Comparable accommodation options were used in the Bell Model 47 series. Production of the Model 12 series ended in 1965 when over 2,000 had been built, and the line was reopened in the early 1980s not only for the support of the many existing Hiller 12s, but also for the production of new aircraft with improved engines and updated avionics to satisfy a developing market for light-weight and relatively cheap helicopters suitable for agricultural work, aerial survey, policing, forestry patrol and firefighting, and executive transport.

Perhaps the type that still epitomizes the light helicopter most redolently for most people is the Bell Model 47, the first helicopter in the world to receive civil certification.

With seating for up to four including the pilot, such helicopters were quite inexpensive to buy and to fly, for a small piston engine developing less than 200 hp was practical, and maintenance of the simple airframe and instruments was not excessive. The other side of the coin, however, was a definite lack of payload and range. This mattered relatively little to the civil operator at the time, but proved a hindrance to the military, who had by 1950 explored the potential roles of the helicopter in considerable depth and decided that much could be achieved by types with greater payload. The problem was quite simple to identify, but at the time virtually impossible to solve: this was the powerplant, which had to be a piston engine of the air-cooled type offering better power-to-weight ratios than similarly rated liquid-cooled inline engines. The small utility helicopter could operate effectively on a single horizontally-opposed engine, but larger helicopters were dependent on air-cooled radials. The trouble with such engines, however, was that their very weight limited the payload, while their bulk combined with their weight to dictate installation as low as possible and also towards the nose for better airflow and cooling. This in turn meant a long transmission shaft running up diagonally from the engine through the fuselage to the gearbox associated with the rotor drive above the cabin roof. Internal volume was thus reduced by the engine and transmission shaft, and payload further reduced by the weight of the transmission shaft. Even if the engine were located directly under the rotor, it still had to be low in the fuselage and drive via a transmission shaft, with consequent weight and volume penalties.

During the late 1940s and early 1950s, therefore, the most useful helicopters were light types designed for limited carriage of passengers or freight, and apart from the types already mentioned, the most successful were the British Bristol Sycamore and Saunders-Roe Skeeter, the French Sud-Ouest SO 1221 Djinn and the Soviet Mil Mi-1, the last being given the NATO reporting name 'Hare'. These were all built in what were by the standards of the day useful numbers, the Djinn being the most fascinating of the series as it used the 'cold jet' propulsion system. This used a Turboméca Palouste gas turbine as a generator of compressed air, which was then piped out to nozzles at the tips of the rotor blades to drive the rotor through torqueless jet propulsion.

But heavier helicopters were already on their way, building on experience with types such as the

***Above**: Rudimentary it might be, but experiments during the 1950s with slung loads (such as the pair of oil drums seen here under a Bristol Sycamore) paved the way for today's utility and flying crane helicopters, which can operate with the most diverse and heavy assortment of slung loads.*

***Below**: Though seen in the markings of British European Airways, the Bristol Sycamore Mk 3A was operated only on a trial basis by this civil operator, this limited exercise confirming that the piston-engined helicopter was of very limited commercial use in the passenger-transport market.*

Sikorsky S-51, whose design was originally designated VS-372 in 1943 when the type was ordered by the US Army Air Force as the R-5. The object was to provide a larger and more capable observation helicopter than the R-4/R-6 series, this being achieved by the use of a new type of fuselage with a crew of two seated in tandem to permit a smaller cross section, a reversed tricycle landing gear, and a powerplant comprising one 450-hp Pratt & Whitney R-985 Wasp Junior radial engine. The

S-51 series was extensively developed as a dual-control trainer, air-sea rescue helicopter with a side-mounted winch, and four-passenger transport, the last having the standard tricycle landing gear arrangement introduced on the R-5D version, which also introduced the R-1340 radial used on this and later military models.The S-51 series was also used by the US Navy with the designation HO3S, and the type was built under licence in the UK with the 525-hp Alvis Leonides radial as the Westland Dragonfly. Earlier Sikorsky helicopters had been used in the UK as the Hoverfly Mk I (R-4 series) and Hoverfly Mk II (R-6 series). Indications that performance and standards were going up were, of course, items such as the need for improved streamlining, but also the introduction of sturdier structures with features such as metal- rather than fabric-covered rotor blades.

The process started with the S-51, was continued with the considerably more advanced Sikorsky S-55 series, a classic helicopter which remained in production with Sikorsky and three licensees (Westland, Mitsubishi and Sud-Est) for more than 10 years for a total of more than 1,750. This production quantity, the very ubiquity of the type, and its survival as an operational type in civil and military hands right up to the present day were all indicative of the success of the S-55. The initial version of the S-55 was the YH-19 evaluation prototype for the US Air Force, first flown in November 1949 on the power of a 600-hp Pratt & Whitney R-1340 radial engine. Key to the design was a larger main rotor, the relocation of the engine to the nose of the fuselage in comparison with the S-51's position under the rotor, and the incorporation of a large cabin supported on quadricycle landing gear. The tail rotor was supported at the end of a long tail boom that was angled down slightly and braced with a triangular fillet to the rear of the cabin compartment, a feature that became almost a trademark of Sikorsky piston-engined helicopters. Although not a spectacular load-lifter by current standards, the S-55 was a considerable improvement on its predecessors, and entered service with the US Air Force and US Army as the H-19 Chickasaw, with the US Navy as the HO4S and with the US Marine Corps as the HRS. Most of these models were dedicated to personnel transport, but the US Navy's experience with the type led it to the conclusion that in the HO4S it had the first makings of an anti-submarine weapon. Evaluation of earlier types had shown that the helicopter was an ideal platform for such a task: it could operate from relatively small escort vessels accompanying the convoys which would be the attacking submarines' main target, and could then transit from

Right*: Though obviously staged, this photograph of the first Sikorsky S-55 to enter service with the US military (as the YH-12 of the US Air Force) indicates the growth of the payload during the 1940s, this helicopter of 1950 being capable of carrying an eight-man squad.*

Below*: The Sikorsky S-51 series was built in the UK as the Westland Dragonfly, the type being used for the validation of maritime air-sea rescue.*

Bottom right*: Seen in company with Westland Whirlwind and Wessex single-rotor helicopters, the Westland Belvedere was designed by Bristol, and was only moderately successful in introducing a twin-rotor configuration.*

the parent ship to the area in which a submarine was suspected in considerably less time than was possible for a destroyer or frigate. Once in the target area it could search for the submarine, covering large areas or hovering as required, and then return to the convoy which had not meanwhile been deprived of its close escort of anti-submarine ships. The one fly in the ointment was the helicopters' lack of payload: the HO4S could carry sensors to detect the submarine, and it could carry weapons to destroy the submarine; unfortunately it could not do both at the same time, so the US Navy decided that what was needed was a hunter-killer team of two helicopters, one with sensors and the other with weapons, or alternatively a hunter-killer team of sensor-equipped helicopters which could then call in a destroyer for the actual 'kill' of the submarine. Experience with the HO4S thus led the US Navy towards the tactical anti-submarine doctrine that still prevails today. The Royal Navy, on the other hand, saw the helicopter, even in the limited form represented by the Westland-built Whirlwind with 750-hp Alvis Leonides Major radial, as an ideal single-vehicle hunter-killer of submarine: in the Whirlwind HAS.Mk 7, for example, the cabin was occupied by a tactical crew to man the radar and dipping sonar, and a ventral weapons bay contained a lightweight acoustic-homing torpedo. Performance of the heavily laden HO4S and Whirlwind series was at best marginal, but much valuable operational experience was gained by the US Navy and Royal Navy.

As a troop transport the H-19 series carried up to 10 troops, with an alternative configuration for six litters and one medical attendant possible, while the not insignificant number of sales of the S-55 series on the civil market were for versions generally tailored to eight passengers carried in a fair degree of comfort. The Whirlwind, it should be noted, was largely responsible for the development of helicopter search-and-rescue in coastal areas, the Royal Air Force operating the type in a vivid yellow colour scheme to so great an effect that the concept of SAR became commonplace in

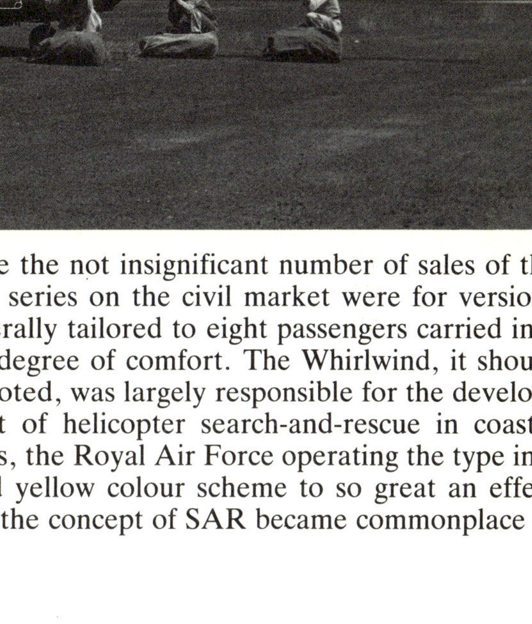

a short time and so hastened public acceptance of the helicopter.

Successful as the S-55 undoubtedly was, the US services in particular needed a machine with heavier lift capability. This specialist requirement was met partially by the extraordinary Sikorsky S-56, which entered military service with the US Marine Corps as the HR2S and with the US Army as the H-37 Mojave. The type resulted from a US Marine Corps' requirement for a helicopter to perform the assault transport role with a load of 26-fully equipped troops. What was needed here was good performance, twin-engine powerplant for increased survivability, and an unobstructed cabin for the accommodation of payload. The resultant S-56 had two engines located in external nacelles but geared to drive a single main rotor, clamshell nose doors allowing entry into and exit from a large cabin, and retractable landing gear. There was also a winch able to handle loads of up to 2,000 lb. Ordered in May 1951, the first S-56 prototype flew in December 1953, and was immediately recognized as a machine of special capabilities, even if it was limited by the bulk and weight of its two massive 1,900-hp Pratt & Whitney R-2800 Double Wasp radials. Even so, it was the Western world's largest and fastest helicopter for some time, and held two payload-to-height records between 1956 and 1959. Deliveries began in 1956, but the specialized nature of the S-56 meant that orders were relatively small. One most interesting experiment with such a helicopter was the installation of AN/APS-20E search radar under the nose in a bulbous radome in two HR2S aircraft, the machines being used as trials aircraft for the possible development of helicopters as airborne radar pickets for important naval task forces. As always with such piston-engined machines, the limitation that made the experiment impractical was the small fuel load that could be

***Right**: The Fairey Rotodyne had prodigious performance, including a cruise speed of 185 mph, but was cancelled by the British government in 1962.*

***Below**: The Sikorsky H-37 Mojave was the helicopter giant of its day, with two radial engines in nacelles located one each side of the fuselage to drive a single large main rotor without obstructing the payload bay within the central fuselage.*

carried when the massive radar equipment and associated crew were installed.

Altogether more successful as a piston-engined helicopter was the next offering from Sikorsky, the S-58. This design originated from a 1952 US Navy requirement for a larger and more capable helicopter to replace the HO4S (S-55) as an anti-submarine platform, and first flew in XHSS-1 prototype form in March 1954. Although obviously related to the S-55, the S-58 featured a much sleeker fuselage with the pod-and-boom configuration of the S-55 replaced by a type more closely resembling that of an aircraft, yet with basically the same arrangement of powerplant, cockpit and rotor system. The landing gear was of the tailwheel type, with shorter main unit legs and a retractable tail wheel. The powerplant was a 1,525-hp Wright R-1820 radial, which increased payload and/or range appreciably. However, as the US Navy soon appreciated, the type was still capable of carrying either sensors or weapons over a tactical range, but not both, so the navy was back almost where it had begun with the HSS-1 Seabat. But the success of the S-58 series was assured by its capabilities as a 'conventional' military and civil transport, the large cabin being able to accommodate 10 fully-equipped troops (12 in later versions) or 12 passengers in airline configuration. The military variants were the H-34 Choctaw for the US Army and the HUS Seahorse for the US Marine Corps. The type was widely used for utility transport, and special versions were developed for tasks such as the recovering of spacecraft after they had splashed down at sea, and as the lavishly-equipped Presidential transport. The type was also used by the military and by civil operators as a flying crane, with a substantial load slung under the fuselage for location in otherwise inaccessible sites. Production of the S-58 series reached 1,281 by Sikorsky, with others built under licence by Sud-Aviation in France. And as we shall see below, the S-58 was radically transformed by the design expertise of the British Westland company.

Helicopters of the size and capability of the S-58 series were the almost exclusive preserve of Sikorsky among the Western nations, but in the USSR the Mil design bureau also came up with a comparable design, the phenomenally successful Mi-4, allocated the NATO reporting name 'Hound'. This first flew in August 1952, and though resembling the S-55 in appearance, was more comparable with the S-58 series in performance and capabilities. Powered by a 1,700-hp Shvetsov ASh-82V radial, the Mi-4 lacked any of the S-58's pugnacious grace, but looked and was an extremely workmanlike design tailored specifically to the requirements of the Soviet air arm for a trooping and assault helicopter. To this end the rear of the cabin was fitted with clamshell access/exit doors for the load which could comprise 14 fully-equipped troops, or a GAZ command vehicle, or a 76-mm anti-tank gun, or two motorcycle

Below: *The first tactically important helicopter developed by the Soviets was the Mil Mi-4, named 'Hound' by the NATO Air Standards Co-ordinating Committee for easy identification purposes. The Mi-4 resembled the Sikorsky S-55 in layout, but was closer to the S-58 in size and performance. The type was built in very large numbers for domestic use and for export.*

Below right: *A Sikorsky HUS-1 Seahorse of the US Marine Corps demonstrates the type's capability for rapid rescue as it picks up Alan B. Shepherd after the latter's suborbital flight in a Mercury spacecraft on 5 May 1961. Helicopters were to become wholly familar appendages to spacecraft splashdown before the emergence of the space shuttle orbiters in the 1980s.*

Small and dainty, the Hughes Model 269A still offers good performance thanks to its excellent basic design. Seen here is an example of the type in US Army service as a TH-55A Osage dual-control pilot trainer, a particularly important type in times of rapid expansion.

combinations, or up to 3,527 lb of assorted freight. The type was rushed into widespread service during 1953, and was soon available in the thousands.

The capability of the Mi-4 was soon translated into the naval sphere, where a specialized anti-submarine model was developed with search radar in a radome under the nose, a magnetic anomaly detection set with a towed 'bird' for the sensor, and a variety of electronic aids including sonobuoys. And apart from its military roles, the Mi-4 was developed for passenger roles as the Mi-4P and as an agricultural aircraft with the designation Mi-4S. As a passenger machine the Mi-4P could carry up to 11 passengers with a fair degree of comfort, or 16 passengers in a high-density configuration with comfort relegated to a secondary position. Alternatively, the Mi-4 could be used as an ambulance helicopter with accommodation for eight litters and a medical attendant. Visually, the Mi-4P is distinguishable from the Mi-4 by its square rather than round cabin windows, by its lack of wheel fairings, and by the deletion of the ventral fairing that in the military model houses the navigator and/or extra fuel. The Mi-4S was developed for agricultural use, with the ability to carry a hopper for 2,205 lb of chemical dust or a tank for 352 Imp gal of pesticide or firefighting fluid. Other roles were aerial survey, search-and-rescue, and ice patrol. Apart from extensive production in the USSR, the Mi-4 was also widely built in China.

These heavyweight helicopters were partnered in service by lightweights such as the Bell 47, Hiller 12 and Mi-1, and also by the considerably more advanced and delightful Hughes Model 200 plus its derivatives. The Model 200 design resulted from studies during the early 1950s by the Hughes Tool Company into future needs in the light helicopter category. These studies revealed that current types did not obtain the best performance possible because of excessive structure weight, which required additional power and so made the machine not only clumsier in the air but more expensive to buy. So Hughes worked decisively towards the real lightweight helicopter, with no structural frills to reduce performance, which on the power of a 170-hp Lycoming was shown to be sparkling when the Model 269 prototype started flight trials in October 1956. Alterations before the type was certificated in April 1959 were restrained mostly to the tail boom, which became a lengthened tubular structure with a small tailplane surface on the starboard side only. The performance of the type had already commended itself to the US Army, which was in the throes of an enormous expansion in helicopter strength, and was ordered as the TH-55 Osage dual-control trainer. In July 1960 Hughes decided to press

ahead with civil production of two basic variants, the Model 269A two-seater and the Model 269B three-seater with a slightly enlarged cabin. The two-seater was subsequently redesignated just Model 269, while the three-seater became Model 369. Hughes has since gone on to the design of other helicopters, but has in general abided with the basic lightweight design philosophy that led to the Model 269, so ensuring maximum performance from the chosen powerplant.

An advanced design featuring the 'fenestron' buried tail rotor, the Aérospatiale SA 341 is in widespread service, but has proved to have severe operational limitations. Seen here are Gazelle HT.Mk 2s of No. 705 Squadron. Fleet Air Arm, based at RNAS Culdrose in the 1970s.

Chapter 3
Enter the Turbine

But by the time the Hughes Model 269 was under development, a radical innovation had altered the face of rotary-wing flight. This was the introduction of turbine power, which at a stroke removed many of the limitations imposed upon earlier helicopters by the piston-engine powerplant. Compared with a piston engine of comparable power, the turbine is lighter and smaller, requires

Right: *The Aérospatiale Alouette II was the first turbine-powered helicopter to enter widespread service, and soon confirmed the advantages of the powerplant for reliability and improved utility in roles such as air-sea rescue.*

Below: *In maritime applications the Alouette III can carry a comprehensive survival suite and also a wide variety of operational equipment ranging from sensors such as radar and sonar to weapons such as depth charges and lightweight homing torpedoes.*

Above: *More capacious and better streamlined, the Aérospatiale SA 316B Alouette III offered considerable advantages, and was also developed as a battlefield helicopter with powerful armament, this including up to four SS.11 anti-tank missiles on outriggers.*

less fuel, is mechanically simpler and is more reliable and, quite importantly, produces less vibration as its parts rotate rather than reciprocate. This meant that the engine could be removed from its position low in the fuselage and moved up to a location adjacent to the rotor shaft and gearbox above the cabin, so lightening the installation further by the removal of the need for a heavy transmission shaft. This also left the fuselage clear for the extra payload made possible by the lighter powerplant weight, and ensured that the crew and passengers enjoyed a more comfortable flight without the heavy vibration associated with the earlier piston-engine powerplants. In a way, this adoption of turbine power for the helicopter was as important and far-reaching as the adoption of nuclear power for submarines, which became true underwater vehicles rather than submersibles.

The world's first production helicopter with a turboshaft powerplant was the Sud-Aviation SE 313B Alouette II. This derived from the SE 3101 piston-engined single-seater first flown in June 1958 on an 85-hp Mathis engine and developed into the SE 3110 two-seater with a 200-hp Salmson engine. Production of the SE 3110 was planned, but then the development of turbine engines had reached a sufficiently advanced state for Sud-Aviation to consider a total alteration of the type by the adoption of the 360-hp Turboméca Artouste turboshaft in a prototype designated SE 3130 Alouette II. This first flew in March 1955 and soon revealed its phenomenally improved performance, including a world helicopter altitude record of 26,932 ft. The type was placed in production to meet a growing stream of military and civil orders, and the revised designation SE 313B was allocated in 1967. The extra payload was naturally appreciated by civil operators, but proved most useful to military users of the Alouette II, which was soon to be seen with a variety of external weaponry ranging from a pair of AS.11 or AS.12 anti-tank missiles, rocket pods, machine-guns, cannon and also torpedoes for the light anti-submarine role. Other equipment applications were pioneered for humanitarian and civil roles, including sling equipment for the carriage of a 1,102-lb external load, a winch for the rescue of survivors (in both mountain and maritime disasters), and external panniers attached to the landing gear for the carriage of two litters.

Sud had already seen that the application of the improved Turboméca Astazou would further enhance the Alouette II's already formidable capabilities, and the prototype of this the SA 318C Alouette II Astazou, first flew in January 1961 with a 530-shp Astazou IIA turboshaft derated to 360 shp for greater reliability and longer life. The Alouette II's Artouste IIC6 had been similarly derated from 530 shp to 360 shp, but the Astazou offered reduced fuel consumption and was, in the SA 318C, combined with the improved and strengthened transmission developed for the Alouette III. Another hybrid, developed to the specific requirement of the Indian Army for a versatile helicopter with good performance at high altitudes was the SA 315B Lama. This was in essence the airframe of the Alouette II fitted with an 870-shp Artouste IIIB turboshaft derated to 550 shp, and the transmission and rotor system of the Alouette III. The success of combination is attested by the establishment of a new world helicopter altitude record in June 1972, when a Lama assembled by Hindustan Aeronautics Ltd reached 40,815 ft.

Operational experience with the Alouette II had quickly convinced customers and manufacturer alike that the basic type could be profitably and easily developed into a more capacious type, which emerged in February 1959 as the SE 3160 Alouette III prototype, powered by an 870-shp Artouste IIIB derated to 550 shp, fitted with a transmission and rotor system of greater strength to absorb the extra power, provided with an enclosed semi-monocoque tail boom in place of the Alouette II's open lattice-work type, and enlarged in the cabin area to accommodate a total of seven persons including the pilot. The type was an immediate success, and entered large-scale production in 1961 against civil and military orders that eventually went well over the 1,000 mark. The initial SE 3160 series was replaced in production during 1970 by the improved SA 316B, which in turn gave way to the SA 316C (with 870-shp Artouste IIID derated to 600 shp) in 1972. The type was also transformed in the same way as the Alouette III by the adoption of the more economical Astazou turboshaft to produce the SA 319B Alouette III Astazou with the 870-shp Astazou XIV derated to 600 shp. The success of the two basic Alouette models is proved not only by their large production totals, but also by the extent to which they were exported. They were thus instrumental in developing the notion of the helicopter's utility in corners of the world that had hitherto remained immune from rotary-wing flight, and though part of the types' success can be attributed to sound initial design, part must also be attributed to the turboshaft powerplant, which soon put the lie to worriers who claimed that turbine power was effective only in the hands of technologically advanced nations with the skilled manpower to maintain these new and advanced engines. Operations soon proved that the turboshaft was inherently more reliable than the piston engine, and that maintenance was no problem in any event. And like the Alouette II, the Alouette III in both its powerplant variants proved an admirable multi-role helicopter for civil and military applications, continuing and improving upon the trends pioneered with the Alouette II series. Perhaps the most important of these developments was in the military sphere, where the Alouette III capitalized on experience with the Alouette II to evolve a whole series of tactical uses for the helicopter as an armed battlefield helicopter, able to transport men and so provide greater battlefield mobility, but also able to tackle enemy movements, tanks and strongpoints with missile, rocket and gun armament, using its agility and hovering capability to keep out of trouble. As a flying crane the Alouette III can carry a slung load of 1,653 lb, and as an air ambulance it can accommodate in the cabin two sitting and two lying casualties.

The turbine revolution quickly swept through the helicopter-manufacturing industry. In the UK Westland immediately applied the new type of powerplant to the S-58, licence rights to which were acquired in 1956. The first example received by Westland was a standard HSS-1 with radial powerplant, but by May 1957 it had been transformed with a Napier Gazelle turboshaft of 1,100 shp. Trials were highly successful, and the revised type was ordered into production as the Westland Wessex utility and anti-submarine helicopter with the 1,450-shp Gazelle Mk 161 turboshaft. Even though this developed some 75 hp less than the radial used in the HSS-1, performance was improved by a considerable margin thanks to the turboshaft's advantages mentioned above. The Wessex anti-submarine variant could alternatively accommodate 16 passengers or eight litters and one medical attendant, and the slung load capability was 4,000lb. Later developments to the utility line produced Wessex helicopters with two Rolls-Royce Gnome turboshafts in place of the single Gazelle, with important results such as extra reliability in the event of an engine failure.

The Whirlwind developed from the S-55 was also adapted for turbine power in the Whirlwind

__Below:__ The Westland Wessex was a turbine-powered derivative of the S-58 series, and widely developed for anti-submarine warfare, search and rescue, and tactical transport.

15
23
S.A.R.

CU
DANGER
ROYAL NAVY

The Falklands campaign of 1982 showed how dangerous was the Royal Navy's lack of an airborne early warning aircraft, a deficiency now being made good by an adaptation of the Sea King HAS.Mk 5 with a swivelling lateral randome for the antenna of the advanced air-search radar equipment.

Above: Developed specifically for anti-submarine warfare, the Sikorsky SH-3A Sea King was powered by twin turbines for safety.
Above left: *Built in Italy as the Agusta AS-61R, this helicopter is equivalent to the US Coast Guard's HH-3F Pelican SAR 'chopper'.*

Series 3 model, which after trials with twin Blackburn Turmo engines settled on a powerplant of one Gnome to produce the definitive search-and-rescue variant, the Whirlwind HAR.Mk 10. After the success of the turbine power-versions of the S-55 and S-58 had been confirmed, many civil and military operators had their piston-engined aircraft modified by Sikorsky or other specialist companies to turbine power.
weapons, but not both. The airframe was capable of accepting both, and the adoption of a turbine powerplant promised to improve power-to-weight ratios to a degree that it would be feasible to do so. So was born the S-61 series, whose prototype first flew as the XHSS-2 in March 1959. This was powered by two General Electric T58 turboshafts in the favoured position above the cabin for maximum usable volume and the extra reliability of a twin-engine powerplant for overwater operations. At the same time the XHSS-2 was designed with a boat hull and lateral stabilizing sponsons into which the main units of the tailwheel landing gear retracted. Though not intended to operate from water, this sensible capability was built into the S-61 design and a safety feature and confidence-builder for crews. The XHSS-2 was clearly a winner, and was ordered into production as the SH-3A Sea King with dunking sonar and provision

Sikorsky was also at work on a turbine-powered type in response to a US Navy requirement of 1957 for an anti-submarine helicopter more capable than the HSS-1 series. As noted above, the limitation of the S-55 series in the anti-submarine role had been the ability to carry sensors or for some 840lb of weapons. However, as the US Navy was now firmly established in the tactical doctrine of using helicopters for the hunting of submarines and surface warships for their killing, the SH-3A was mostly used in this single role. Further development produced the SH-3D improved anti-submarine helicopter with 1,400-shp T58-10 engines, the SH-3G utility helicopter with removable anti-submarine gear, and the SH-3H multi-role helicopter with improved anti-submarine equipment and undernose radar for the detection of incoming sea-skimmer missiles.

These four naval variants were produced by Sikorsky under the company designation S-61A, which also covered 31-seat troop transports for Malaysia, search-and-rescue aircraft for Denmark and such specialist versions as RH-3 mine countermeasures, VH-3 VIP transport and CH-3 utility transport versions. All along, however, Sikorsky had appreciated that the basic design was amenable to a wide diversity of applications, and the S-61 was also developed as the S-61R military transport, a use pioneered in American service by the CH-3Bs borrowed from the US Navy by the US Air Force. The success of this simple conver-

Used for a multitude of company-sponsored projects, this S-61A epitomizes the compact twin-turbine design that makes the Sea King a classic of its type.

sion added impetus to the S-61R programme, which first bore fruit in the form of the CH-3C, a significantly altered helicopter with a different fuseage fuselage characterized by rear ramp doors strong enough to allow the onloading of small vehicles, pieces of light artillery and the like, loading being facilitated by the incorporation of kneeling landing gear to accommodate the helicopter to the height of any truck that might be offloading into it. The initial production CH-3C had 1,300-shp T58-1 engines, but the CH-3E had 1,500-shp T58-5s. Modifications were the HH-3E Jolly Green Giant combat-area rescue helicopter with inflight-refuelling capability, armour and armament, and the HH-3F Pelican advanced rescue helicopter for the US Coast Guard. Both the HH-3 variants have made great names for themselves, the HH-3E becoming almost legen-

dary for its success in the Vietnam War.

Sikorsky also appreciated the S-61's civil potential, producing the type in two basic types, the S-61L and the S-61N. The former is a dedicated landplane derivative without the boat hull of the SH-3 series and with fixed landing gear in place of the combined sponson/retractable landing gear of the SH-3. A crew of three is carried, and payload comprises 28 passengers seated in airliner comfort. The type first flew in December 1960, and enjoyed modest sales success. Greater commercial advantage came to the company from the S-61N, which first flew in August 1962. This model retained the amphibious capability of the SH-3 series, and was geared to the needs of operators servicing the oilfields rapidly being exploited in offshore waters. Maximum passenger capacity is 30.

The Sikorsky licensee in the UK is Westland, and this latter company has also developed the Sea King design to suit particular British requirements. There has thus evolved the truly effective hunter-killer Sea King anti-submarine helicopter developed along the lines of the Wessex with a tactical compartment for a sensor operator and tactical co-ordinator, a two-man crew fully capable of detecting a submerged submarine without recourse to a warship. Westland has also developed the Sea King into one of the world's finest air-sea rescue helicopters, and has produced the Commando battlefield helicopter based on the airframe and dynamic system of the Sea King but fitted with fixed landing gear.

The development of turbine-powered helicopters has not, of course, been a Western preserve, for the USSR quickly stepped onto this technical bandwagon. The first production helicopter of Soviet origins with a turboshaft powerplant was the Mil Mi-2 'Hoplite', basically a version of the Mi-1 recast to accommodate two 435-shp Isotov GTD-350 turboshafts above the cabin, which thus becomes available for payload. An indication of the turbine's effect on payload and performance is given by relative figures for the Mi-1 and Mi-2: whereas the Mi-1 could carry a pilot and three passengers, the Mi-2 could uplift a pilot and up to eight passengers, speed and range being 96 mph and 373 miles for the Mi-1, and 130 mph and 105 miles with a heavier payload for the Mi-2. However, far more impressive Soviet turbine-powered helicopters were already under development, the two most important of these early series being the Mi-6 'Hook' and Mi-8 'Hip', developed for heavy- and medium-lift duties respectively.

The Mi-6 first flew in 1957, and remained for some 10 years the world's largest helicopter. A clean design of pleasing lines, the massive Mi-6 was powered by two 5,500-shp Soloviev D-25V turboshafts, and entered production during 1960.

Left: *The versatile Westland Sea King family is seen in some of its variants. From right to left these are a company research model, the Sea King HAS.Mk 1 baseline variant, the Sea King HAR.Mk 3 SAR model, the Sea King HC.Mk 4 tactical model and the Sea King HAS.Mk. 5 definitive anti-submarine version.*

Top: *Derived from the Sea King with fixed tailwheel landing gear, the Westland Commando is a multi-role tactical helicopter capable of accepting a diversity of weapon fits.*

The type was designed principally for the opening up of areas such as Siberia, where there were few facilities for fixed-wing aircraft of comparable capability, and the Mi-6 soon proved itself an admirable type, being capable of lifting a 44,350-lb load, and on one occasion lifting a 22,046-lb load to 16,027 ft. These were exceptional loads, but the standard Mi-6 in service with Aeroflot could carry an internal payload of 26,455 lb including 65 passengers (120 in a high-density arrangement) or 41 litters and two medical attendants. A specialized flying crane was also developed from the Mi-6 in the form of the Mi-10 'Harke', which features a redesigned fuselage comprising a long boom with the cockpit at the front, powerplant and rotor system in the centre and tail rotor at the rear. The boom has a flat undersurface against which special payload pods can be hoisted, and the type is fitted with straddling quadricycle landing gear to permit the fitting of outsize loads. Unlike the Mi-6, which uses wings to develop some 20 per cent of the lift in forward flight and so offload the massive rotor, the Mi-10 has no wings, being intended for short trips with substantial loads. In May 1965 the Mi-10 set a world helicopter payload-to-altitude record, lifting a load of 55,347 lb to a height of 9,318 ft. A derivative of the Mi-10, which is designed specifically for loads that need not be straddled, is the Mi-10K with shorter landing gear and two 6,500-shp turboshafts, designed for the movement of slung loads weighing as much as 30,865 lb.

The Mi-8 appeared in 1961, and may be regarded as the turbine-powered version of the Mi-4 in much the same way as the Mi-2 is the turbine-powered version of the Mi-1. The prototype Mi-8 was powered by a single 2,700-shp Soloviev turboshaft, but it was soon appreciated that a twin-turbine powerplant would offer greater reliability, and the type was revised with two 1,500-shp Isotov turbines and a five-blade main rotor in place of the original four-blade unit. Entering service with the Soviet forces in about 1967, the Mi-8 is still in large-scale production for civil and military operators, and is certainly the single most important type in the Communist bloc's helicopter inventory. The type is used for the whole range of helicopter tasks, ranging from civil transport with 28 passengers or 8,818 lb of internal freight, via flying crane with a slung load of 6,614 lb to assault transport with troops and an extremely heavy armament of unguided rockets. The basic type has also been developed into the Mi-14 'Haze' anti-submarine helicopter with a boat hull, and into the uprated and generally improved Mi-17 utility helicopter that appeared in the early 1980s by a combination of the Mi-14's dynamic system with the basic airframe of the Mi-8.

The USA also moved rapidly into the turbine-engined helicopter field for transport purposes. Already mentioned has been the S-61 which

appeared in several transport versions, but perhaps more important are a number of special helicopters from Bell and Boeing Vertol. In 1955 Bell won a US Army competition to decide the new generation of utility helicopter for that major rotary-wing operator. The winning design was the Bell Model 204, evaluated under the military designation XH-40. Initial production version was the HU-1 Iroquois, the letter prefix giving rise to the unofficial but far better known name 'Huey'. In 1962, when the whole designation system used by the US forces was rationalized, the HU-1 became the UH-1, which was produced in vast numbers (several thousands) and in several major variants.

First was the UH-1B, powered by a 960-shp Lycoming T53-5 or 1,100-shp T53-11 turboshaft and able to accommodate, in addition to the crew of two, seven passengers or three litters. The type became operational in Vietnam during 1962, and became the real work horse of the US and allied ground-troop mobility effort during that disastrous war. The type soon began to sprout armament, UH-1Bs being seen with four machine-guns and two pods of unguided rockets. French experience, particularly against guerrilla forces in North Africa, had indicated that the helicopter was a potent weapon for such operations, and the Americans were quick to appreciate this point for themselves in Vietnam: troops could be moved rapidly and in some numbers by helicopter to establish troop concentrations anywhere in the country, and the helicopter was also ideally suited to the close-support of the landed troops with weapons such as machine-guns, unguided rockets and grenades. So was born the concept of airmobile operations with helicopters for movement, fire support, communications relay, resupply, casualty evacuation and airborne command. The UH-1 series was ideally suited to all these tasks, while the more agile and higher-performance Hughes OH-6 Cayuse series (Hughes Model 500) was well suited to reconnaissance and forward air control in conjunction with the 'Hueys'. The one major drawback of the 'Huey' series was found to be its relative slowness, dictated by the broad-chord two-blade main rotor design, which was also relatively noisy and so made difficult the securing of total tactical surprise.

By 1963 the UH-1B was being supplanted by the UH-1D, the most numerous of the series, and developed under the Bell designation Model 205. This resulted from company and military realization that the powerplant and rotor system of the Model 204 was capable of accommodating a

Below left: *Successor to the Mi-4, the Mil Mi-8 'Hip' is a highly capable turbine-powered machine, seen here in the form of a Mi-8P of the Finnish air force.*

Below: *The Mil Mi-17 is an uprated development of the Mi-8, the use of an improved powerplant maintaining the type's utility for many years to come.*

RAAF
RAAF
RAAF
279
RAAF
770
RAAF
379

greater load than could be fitted into the somewhat small fuselage. Thus appeared the Model 205 with a larger fuselage able to accommodate up to 14 troops or six litters and one attendant, and powered by the T53-11. If any of the 'Huey' models can be described as the work horse of the Vietnam war, it must be the UH-1D. The model 205 was also produced for civil use. Another variant that proved important during the Vietnam war was the UH-1C developed from the UH-1B with an improved rotor and the T53-11 engine. This model became available in 1965, and was soon joined by a growing number of other UH-1 variants such as the UH-1E assault transport for the US Marine Corps, the UH-1F missile-site support helicopter for the US Air Force, the

Left: ***The Bell 'Huey' series was the classic light tactical transport of the 1960s and 1970s, and is seen here in the form of UH-1D/H helicopters of the Royal Australian Air Force.***

Right: *Another way to delivery infantry is by rappeling, as from this New Zealand 'Huey' tactical transport.*

Below: *In the Vietnam War the 'Huey' was the US Army's maid of all work, carrying troops into landing zones whose defenders had been 'softened up' by door-mounted machine-guns.*

Ultimate in the design philosophy that emerged with the Model 204 series in the late 1950s, the Bell Model 214ST is a highly versatile civil and military helicopter well suited to operations over water as well as over land. The example seen here is operated by Okanagan Helicopters of Canada for the support of offshore resources-exploitation sites.

OKANAGAN

TH-1F dual-control trainer, the UH-1H uprated version of the UH-1D, the HH-1K air-sea rescue helicopter, the TH-1L dual-control trainer for the US Navy, the UH-1L utility helicopter for the US Navy,and the UH-1M night-fighting experimental helicopter for the US Army. Of these by far the most important was the UH-1H, essentially the UH-1D fitted with the 1,400-shp T53-13 turboshaft. The series has also been built under licence, the most important such source being the

Italian company Agusta, which apart from Model 204B, Model 205 and Model 205A versions identical in all major respects with their American counterparts other than an option for the Rolls-Royce Gnome turboshaft, has also developed the specialized AB.204AS anti-submarine version with dipping sonar and an armament of two Mk 44 lightweight homing torpedoes.

At the specific request of the Canadian forces, the Model 205 was also developed into the Model 212 with the Pratt & Whitney Aircraft of Canada Turbo Twin Pac powerplant, a unit rated at 1,250 shp but able to offer far greater reliability than the standard T53 as it has two turbines coupled by a combining gearbox to drive the single rotor. The type was also ordered for the US forces and is best known by the military designation UH-1N. Further development of the same basic series has produced the tadpole-shaped Model 214, originally to an Imperial Iranian air force requirement as the Model 214A Isfahan, but since supplanted by the potent Model 214 BigLifter utility helicopter for civil applications. Four-blade main rotors have also been developed for some of these models, the Model 212 becoming the Model 412 by the adoption of such a rotor.

Agusta also builds the Models 212 and 214 under licence, its two most important derivatives being the AB.212 ASW, an advanced type capable of anti-submarine operations with its dipping sonar and torpedo armament, of missile-detection operations with its search radar, and of missile-direction with the aid of data-link equipment, and the AB.412 Grifone, an multi-capable military utility helicopter.

The Model 206 has proved an immense success in the civil market, where it has been marketed under the name JetRanger in three basic versions, plus the LongRanger 'stretched' version and the TexasRanger export military derivative. Also available on the civil market is the Twin Two-Twelve, the commercial equivalent of the UH-1N/Model 212 series. Bell's current helicopter line is completed by the Model 222, an advanced twin-engined type with retractable landing gear and intended for offshore operations (in support of the energy-exploitation industry) and as a comfortable executive transport. The US military designation for the Model 206 is OH-58A Kiowa.

Mention has been made above of the Hughes Model 500, the beautifully streamlined high-performance successor to the Model 269/369 series. This first flew in February 1963 as the Hughes competitor in the US Army's Light Observation Helicopter competition, the other two finalists being the Bell Model 205 and the Fairchild Hiller FH-1100. The Hughes helicopter, powered by a 317-shp Allison T63, walked away with the honours and was ordered into massive production as the OH-6A Cayuse. However, production slippages and escalating costs led to a reopening of the contest in 1967, when the Bell Model 206A was judged the winner and ordered as the OH-58A. However, no one could fault the superlative performance and technical reliability of the Hughes helicopter, which has been extensively developed for the civil market as the Model 500 and long-nose Model 530, and for the export military market as the Model 500 Defender. In this latter capacity the Model 500 has proved most successful, being ordered in largish quantities by many countries for its performance, relative cheapness and excellent weapons capability. This latter can include anti-tank missiles, heavy cannon, unguided rockets and torpedoes according to role, and the mission capability of the Model 500 Defender series is matched by a high degree of mission specialization in electronics.

At the other end of the size scale are the two twin-rotor designs from Boeing Vertol, namely the Model 107 and the Model 114. The Americans have long realized the potential of the twin-rotor helicopter for heavy-lift duties, the most practical approach being deemed that with interconnected

Left: *Small, agile, and fast, the Hughes Model 500MD Defender packs a mighty punch in the form of four Hughes TOW anti-tank missiles, whose targets can be acquired via the mast-mounted sight while the helicopter itself remains relatively safe under cover.*

Right: *Keynotes of the Hughes design philosophy are maximum power combined with a small but highly refined airframe of great strength, and advanced dynamic components such as the rotor.*

rotors at each end of a long fuselage. The Soviets attempted a comparable programme with machines such as the Yakovlev Yak-24, but problems with this configuration persuaded them instead to try the side-by-side twin-rotor approach in machines such as the Kamov Vintokryl and Mil V-12 'Homer', experimental types with massive rotors at the tips of substantial outriggers in the fashion of the Fw 61. Both these machines had prodigious performance and payload capability, but suffered from severe control problems and did not progress past the prototype stage, Soviet heavy-lift capability being vested in the Mi-6 and Mi-10 series, currently being upgraded with the excellent Mi-26 'Halo' single-rotor helicopter.

The USA, on the other hand, persevered with the twin-rotor type through the Bell HSL anti-submarine helicopter and a series of Piasecki utility and transport helicopters before arriving at the Vertol Model 107, which entered production with two T53 turboshafts interconnected to drive two large rotors, one above the cockpit and the other on a taller pylon at the tail. The type sold moderately well as the Vertol 107 for air-taxi and commuter operations in the hands of some American airlines, but found its mark at the CH-46 Sea Knight assault transport for the US Marine Corps. The Model 107 prototype first flew in April 1958, and the CH-46A was ordered in February 1961 as a transport for 25 troops or 4,000 lb of freight. Successive variants appeared, the definitive production model being the CH-46D with 1,400-shp General Electric T58-10 turboshafts and the ability to carry a slung load of 10,000 lb. From this has been developed by modification the ultimate CH-46F with improved electronics and other detail modifications.

Altogether more capable, however, is the classic Model 114, developed for military service as

***Below:** The introduction of twin-turbine powerplants permitted the development of safe and capable twin-rotor designs such as the Vertol Model 107, used by companies such as Pan Am for urban links with major airports.*

***Right:** Evolved from the Model 107, the Boeing Vertol CH-46F of the US Marine Corps is a capable assault transport for 25 troops.*

***Below right:** The Kamov Ka-22 Vintokryl was an experimental but vast convertiplane weighing over 74,250 lb and being very fast.*

the CH-47 Chinook. This has the same basic configuration as the CH-46, and first flew in prototype form during September 1961 with a pair of Lycoming T55 turboshafts. The initial production model was the CH-47A Chinook, powered by two 2,200-shp T55-5s, and the type soon proved its worth in Vietnam, where it performed miracles of resupply and equipment rescue. The latest production model is the CH-47D, all surviving older models being remanufactured to this standard with 4,500-shp T55-712 turboshafts. Some indication of the type's capability is indicated by an internal capacity for 44 fully-equipped troops or 14,322 lb of freight. However, it is with a slung load that the Chinook is most valuable, the use of three-point suspension allowing the utmost flexibility in loads up to a weight of 28,000 lb including aircraft, light vehicles, pieces of artillery and a variety of rubber fuel tanks. The Model 114 has been further developed as the Model 234 commercial Chinook with a luxurious cabin offering airliner comfort and safety for 44 passengers carried over long stages, in roles such as the support of offshore oil and gas rigs.

The Model 107 is now built only in Japan as the Kawasaki KV-107II for a variety of military and civil tasks such as firefighting, air ambulance work and police patrol, while the Model 114 is built in Italy by Meridionali, an Agusta subsidiary.

Sikorsky has not been left behind in the development of heavy helicopters, its two main offerings being the S-64 and S-65, which share a common dynamic system. The S-64 was produced as a specialized flying-crane helicopter, its prototype flying in May 1962. Few civil orders were forthcoming, and the principal production variant was thus the CH-54A Tarhe for the US Army. This is powered by two 4,500-shp Pratt & Whitney T73-1 turboshafts, and can carry a slung load of 20,000 lb. A development was the CH-54B with two 4,800-shp Pratt & Whitney JFTD12-5A turboshafts, the addition of an extra 6,000 shp increasing payload by 5,000 lb to 25,000lb. Though the S-64 was not in itself a great commercial success for Sikorsky, it paved the way for the S-65 heavy assault helicopter, which first flew in prototype form during October 1964. Powered by two 2,850-shp General Electric T64-6 turboshafts, the CH-53A Sea Stallion initial production version was ordered by the US Marine Corps for assault movement of some 37 troops and the evacuation of up to 24 casualties and four medical attendants. The CH-53A was subsequently developed into the CH-53D with two 3,925-shp T64-413 turboshafts, and other variants were the HH-53B and HH-53C

Below left: *The CH-47C is an advanced military counterpart to the civil BV 234 series, with great payload and excellent reliability.*

Below: *The Boeing Vertol BV 234 is seen in a classic location, an offshore drilling rig whose crew are rotated by helicopter.*

Above: *The BV 234 Commerical Chinook offers its passengers all the comforts and amenities of airline transport.*

combat-area rescue helicopters for the US Air Force. A notable feature of all but the earliest CH-53Ds is the incorporation of provision for minesweeping gear, while the RH-53D is a dedicated mineweeping variant powered by two 4,380-shp T64-415 turboshafts and able to operate the US Navy's standard range of mechanical, acoustic, magnetic and magnetic/acoustic sweeping gear trailed out of the rear and controlled by a powerful winch. Even more impressive is the CH-53E Super Stallion, currently the Western world's most powerful helicopter. This is a radical development of the RH-53D with three 4,380-shp T64-416 turboshafts powered a seven- rather than a six-bladed main rotor of larger diameter than that used in the standard CH-53 series.

Nor has Sikorsky forgotten the smaller helicopter, for currently entering widespread service with the US forces in the Sikorsky S-70 series, designed to meet a utility tactical transport requirement from the US Army for a 'Huey' replacement, but rapidly developed as a true multi-role aircraft. The first prototype flew in October 1974, and the type is powered by two General Electric T700 turboshafts. The baseline model is the UH-60A Blackhawk, able to move a full infantry squad of infantry (11 men) or four litters and one medical attendant in the cabin. Other significant models are the SH-60B Seahawk and the HH-60D Night Hawk. The SH-60B is intended as the LAMPS III (Light Airborne Multi-Purpose System Mk III) helicopter to partner the LAMPS I type, the Kaman SH-2F Seasprite, in US Navy service. These two types are the American equivalent of the Italian AB.212ASW, and are designed so that the SH-2F can operate from smaller warships and the SH-60B from vessels of destroyer-size upward.

The Soviet equivalents to the LAMPS I and utility tactical transport types are, in the first case, the Kamov Ka-25 'Hormone' and, in the second case, the Mi-8 already mentioned and also the signally important Mil Mi-26 'Hind'. The Kamov Ka-25, which is being supplemented and eventually replaced by the similar but more advanced

Left: *The Sikorsky CH-53E Super Stallion is a massively capable three-engined helicopter capable of several heavy-duty roles.*

Below: *Seen with a Sea King hovering above it, the Sikorsky SH-60B Seahawk is depicted in prototype form. The type has been selected as the US Navy's Light Airborne Multi-Purpose System Mk III, to be based on destroyers and cruisers for anti-submarine, anti-missile and utility duties.*

Above: *The Mil Mi-24 'Hind-E' and generally similar 'Hind-D' are truly formidable Soviet gunship helicopters with missiles, rocket pods and a powerful gun allied to advanced sensors and a specialist weapons officer in the optimum nose cockpit.*
Right: *Undoubtedly the most formidable attack helicopter in service, the Hughes AH-64A Apache has advanced sensors and devastating armament, but only at the cost of prodigious expense.*
Above right: *An AH-64A unleashes one of the 16 Rockwell Hellfire laser-guided 'fire-and-forget' anti-tank missiles that form the type's primary armament.*

Ka-27 'Helix', is a fascinating type using co-axial twin rotors powered by twin 990-shp Glushenkov GTD-3BM turboshafts for adequate rotor area with minimum overall dimensions for ease of shipboard handling. The Ka-25 appears in three versions, the 'Hormone-A' being the anti-submarine model with dipping sonar, search radar and a considerable weapons load, the 'Hormone-B' a missile-direction model with search radar and a data-link for the mid-course guidance of long-range missiles, and the 'Hormone-C' the search-and-rescue variant. All three can carry up to 12 passengers in the boxlike fuselage, which permits the type to double as an onboard delivery type.

The Mi-24 'Hind' is a truly formidable aircraft, being available in some considerable quantity as a troop transport and attack helicopter. The troop transport versions are the 'Hind-A' and 'Hind-C', able to carry a full section of eight troops in addition to a crew of four, and powered by two 2,200-shp Isotov TV3-117 turboshafts. Apart from its load of infantrymen, the transport versions of the 'Hind' can carry formidable armament under the stub wings, the weapons including guided-missiles for anti-tank use, unguided rockets and bombs. Basically similar, but fitted with a totally revised nose section, is the 'Hind-D' attack helicopter. This carries an underwing armament comparable to that of the transport 'Hinds', but has a four-barrel heavy machine-gun in an under-nose turret, compared with the single-barrel machine-gun of the transport 'Hinds'. But the important fact is the revised forward fuselage, which accommodates a gunner and pilot under separate fighter-type canopies, the gunner being located in the nose with excellent fields of vision and a useful array of sensors and air-data instruments for accurate weapon delivery.

The attack versions of the 'Hind', of which the 'Hind-D' is the most numerous, are the equivalent

to the various gunship versions of the 'Huey' series, produced as a direct result of American experience in Vietnam under the company designation Model 209. The first AH-1G HueyCobra flew in September 1965, and was immediately apparent as the rotor system and powerplant of the Model 204 series married to an extremely slender fuselage (just the width of a man's shoulders) and advanced weapon-delivery system. The forward fuselage accommodated the pilot and gunner (the latter in front) under a long canopy, and the armament comprised an under-nose turret housing a machine-gun and grenade-launcher, and four hardpoints under the stub wings for machine-gun pods, unguided rocket pods and a cannon pod. Deliveries began in June 1967, and the AH-1G proved an immediate success. Since that time development has been constant, the AH-1G leading to the AH-1Q interim anti-tank version with provision for eight TOW anti-tank missiles, the uprated AH-1R and finally the definitive AH-1S with more power, guided-missile capability and many significant improvements, especially to the weapon-aiming and weapon-control systems. Parallel with the HueyCobra, Bell developed a twin-engined Sea-

Cobra series for the US Marine Corps, using the Pratt & Whitney Aircraft of Canada T400 coupled turboshaft for greater reliability. The initial AH-1J was comparable to the AH-1G, but the current AH-1T Improved SeaCobra has a more advanced dynamic system and provision for fuel/air explosives, missiles, rockets and grenade-launchers.

So far only the Americans and the Soviets have produced dedicated gunship helicopters, the latest American vehicle being the potent but also remarkably expensive Hughes AH-64A Apache. This carries a primary armament of one Chain Gun cannon and up to 16 Rockwell Hellfire anti-tank missiles, which home automatically onto laser light reflected from any vehicle 'illuminated' by a friendly target designator. The type also has extraordinarily advanced helmet-mounted sights for flight and accurate weapon-delivery at night and in all weathers, and a host of survival features. But its expense is likely to keep production to a minimum. The only European gunship is the Agusta A 129 currently undergoing trials. This is considerably smaller than American or Soviet gunships. Otherwise the Europeans prefer battlefield helicopters with multi-role capability, examples being the West German PAH-1 developed from the MBB BO 105 utility helicopter, the French Aerospatiale SA 341/342 Gazelle and the British Westland Lynx. All three of these types is available in civil and military forms, their battlefield roles being reconnaissance and light transport, or anti-tank deployment with the installation of missile-launchers and a stabilized sight. The Lynx has also been developed as a highly effective anti-submarine and anti-ship helicopter, and also into the WG 30 high-capacity civil helicopter. The heavyweight partner of the Gazelle is the Aerospatiale SA 330 Puma, and the improved AS 332 Super Puma evolved from the Puma adds further capability.

There are a host of other helicopters both civil and military, and exciting developments abound to keep the helicopter in the forefront of aviation.

Left: *Turbine-powered helicopters have given ground forces a hitherto-unimagined tactical mobility, allowing operations in areas that were previously thought impenetrable to all but the lightest-armed guerrillas. Here French paratroops fan out after a mountaintop landing from an Aérospatiale SA 330L Puma.*

Below: *The UK's most successful indigenously designed helicopter has been the Westland Lynx, a fast and extremely agile type developed in parallel military and naval versions. Here one of eight anti-tank missiles is fired by a British army Lynx fitted with a stabilized sight in the cockpit roof.*

We would particularly like to thank the helicopter companies for their generous help with photographs and unless indicated below each illustration was supplied by the helicopter's manufacturer. For reasons of space alone, some references have been abbreviated as follows:

Crown Copyright (MOD)-MOD
Military Archive & Research Services-MARS

Front cover: Hughes/Boeing/MBB.p1: NATO. Canadian Armed Forces.4-5: MOD.6-9: MARS.12-13: Imperial War Museum. 16: Imperial War Museum. 17: USN.20-21: USAF.30-31: USAF.34-35.30: MARS.46-47: K Niska.55 (bottom): Novosti Press Agency.60 (top left): Department of Defence.